LEARN TO FACTOR IN 8 MINUTES

by

Arlissa Pinkelton

Copyright © 2013 Arlissa Pinkelton

All rights reserved. No part of this book may be reproduced or transmitted in any form or by any means, electronically or mechanically, including photocopying, recording, or by an information storage and retrieval system without permission in writing from the author of this book.

ISBN: 978-1-494-85219-1

Mathematical content edited by:
Tommy Minton
Lauren R. Thomas

Cover Layout and Design by:
Tanika Ault
Legacy Designs by Tanika

Dedication

*To my husband, Warren, and daughter, Gabrielle.
Thank you both for always cheering me on.*

About The Author

Arlissa Pinkelton is a Mechanical Engineer turned Mathematics professor. She lives with her family in central Florida where she teaches Algebra. Arlissa has loved math since childhood and seeks to inspire other students with that same passion.

Foreword

Learn To Factor In 8 Minutes is the first book in a series of eight *"8 Minute"* books on topics in Algebra. These titles are designed to make Algebra easier and quicker to learn.

<u>Upcoming Titles</u>

Learn Radicals in 8 Minutes

Solving & Graphing Linear Equations in 8 Minutes

Learn About Exponents in 8 Minutes

Learn All You Need To Know About Real Numbers in 8 Minutes

Learn To Add, Subtract, Multiply and Divide Polynomials in 8 Minutes

Learn Rational Expressions in 8 Minutes

Learn Inequalities in 8 Minutes

Table of Contents

Introduction……………………………………………..5

Greatest Common Factor……………………….....7

Factoring Out the Greatest Common Factor……..11

Factoring By Grouping………………………....…..14

Factoring Trinomials in the form of
   $x^2 + bx + c$……………….............…..18

Factoring Trinomials using the
   A-C Method ……………………………….23

Factoring Trinomials using the
   Trial-and-Error Method………………………32

Difference of Squares………………………….…..40

Perfect Square Trinomials…………………………44

Sum and Difference of Cubes………………………48

LEARN TO FACTOR IN 8 MINUTES

Introduction

Factoring has a few key elements and processes that can make the skill easy to learn. This book will give you a quick, down-and-dirty look at the basics needed for factoring efficiently and accurately.

8 Sub-topics in factoring

1. Finding the Greatest Common Factor (GCF)

2. Factoring out the GCF

3. Factoring by grouping

4. Factoring Trinomials of the form $x^2 + bx + c$

5. Factoring Trinomials : AC-Method

6. Factoring Trinomials: Trial-and-Error Method

7. Difference of Squares and Perfect Square Trinomials

8. Sum and Difference of Cubes

What is Greatest Common Factor?

Before we answer that question, let's define a factor. **Factors** are numbers or terms you can multiply together to get another number or term. A number or term can have many factors. For example, the factors of 6 are 1, 2, 3 and 6. Each of these numbers can be multiplied by another of these numbers to get 6.

The **greatest common factor (GCF)** is the greatest factor common to all of the terms in a polynomial. To find the GCF, you must think of the terms of the polynomial as products. For example, the GCF of 6 and 8 is 2. The factors of 6 are 1, 2, 3, and 6. The factors of 8 are 1, 2, 4, and 8. The greatest factor that they have in common is 2. Let's find the GCF for $20x^2$ and $15x$. It is $5x$.

Looking at them individually, the factors of $20x^2$ are 1, 2, 4, 5, 10, 20, x, and x^2. The factors of $15x$ are 1, 3, 5, 15, and x. We're looking for the greatest number and variable they have in common. The greatest common number is 5 and the greatest common variable is x. So the Greatest Common Factor (GCF) of that polynomial is $5x$.

You try one. Find the GCF of $18x^3$ and $15x^2$.

Did you come up with $3x^2$? The factors of $18x^3$ are 1, 2, 3, 6, 9, 18, x, x^2, and x^3. The factors of $15x^2$ are 1, 3, 5, 15, x, and x^2. The greatest number and variable they have in common is 3 and x^2. The GCF is $3x^2$.

You can also find the greatest common binomial factors. Of the two terms, $7(x-y)$ and $9(x-y)$, there is a binomial the terms have in common: $(x-y)$ is the greatest common binomial factor.

Here are 8 practice problems for you to complete:

Find the Greatest Common Factor.

1. 24 *and* 36

2. 45, 75, *and* 30

3. 105, 40, 60

4. $7x^3$, $14x^2$, and $21x^4$

5. $15a^4b$, $25a^3b^2$

6. $8c^2d^7e$, $6c^3d^4$

7. $3x(a+b)$ *and* $2y(a+b)$

8. $a(x+2)$ *and* $b(x+2)$

Answers with explanations:

1. GCF: 12

Factors of 24: 1, 2, 3, 4, 6, 8, 12.
Factors of 36: 1, 2, 3, 4, 6, 9, 12, 18, 36

* The greatest number in common is 12.

2. GCF: 15

Factors of 45: 1, 3, 5, 9, 15, 45
Factors of 75: 1, 3, 5, 15, 25, 75
Factors of 30: 1, 2, 3, 5, 6, 10, 15, 30

* The greatest factor in common is 15.

3. GCF: 5

Factors of 105: 1, 3, 5, 7, 15, 21, 35, 105
Factors of 40: 1, 2, 4, 5, 8, 10, 20, 40
Factors of 60: 1, 2, 3, 4, 5, 6, 10, 12, 15, 20, 30, 60

* The greatest factor in common is 5.

4. GCF: $7x^2$

Factors of $7x^3$: $1, 7, x, x^2, x^3$
Factors of $14x^2$: $1, 2, 7, 14, x, x^2$
Factors of $21x^4$: $1, 3, 7, 21, x, x^2, x^3, x^4$

* The greatest factors in common are 7 and $x^2 = 7x^2$

5. GCF: $5a^3b$

Factors of $15a^4b$: $1, 3, 5, 15, a, a^2, a^3, b$
Factors of $25a^3b^2$: $1, 5, 25, a, a^2, a^3, b, b^2$

   * The greatest factors in common are 5, a^3, and $b = 5a^3b$

6. GCF: $2c^2d^4$

Factors of $8c^2d^7e$: 1, 2, 4, 8, c, c^2, d, d^2, d^3, d^4, d^5, d^6, d^7, e

Factors of $6c^3d^4$: 1, 2, 3, 6, c, c^2, c^3, d, d^2, d^3, d^4

* The greatest factors in common are 2, c^2, and $d^4 = 2c^2d^4$

7. GCF: $(a + b)$

Factors of $3x(a + b)$: 3, x, $(a + b)$
Factors of $2y(a + b)$: 2, y, $(a + b)$

* The greatest common binomial factor is $(a + b)$

8. GCF: $(x + 2)$

Factors of $a(x + 2)$: a, $(x + 2)$
Factors of $b(x + 2)$: b, $(x + 2)$

* The greatest common binomial factor is $(x + 2)$

Factoring Out The GCF

Once you've identified the GCF, you can "factor it out" of a term or an expression. Factoring it out is done by dividing the term(s) by the GCF. Let's go back to 6 and 8. Recall the GCF was 2. To factor out the 2, divide 6 by 2. 3 is left. Then divide 8 by 2, which leaves 4. Remember for the $20x^2$ and $15x$, the GCF was $5x$. To factor it out, the $20x^2$ would be divided by $5x$, resulting in $4x$. Then dividing $15x$ by $5x$ results in 3.

Using the expression, $12y^2 + 3y$, let's identify the GCF first. It is $3y$. To factor it out, you divide $\frac{12y^2}{3y}$ resulting in $4y$ and $\frac{3y}{3y}$ leaving 1. Put the GCF on the outside of the parenthesis and the remaining terms inside. The answer is $3y(4y + 1)$.

Another example, $-2y^2 + 3y$, has a negative as the leading coefficient. During factoring, you always want to factor out that negative. So what do you think is the GCF? How about $-y$? After dividing by $-y$ in order to factor it out, you have $-y(2y - 3)$.

Here are 8 practice problems for you to try.

Factor out the GCF.

1. $4x - 20$

2. $6w^2 + 3w$

3. $15y^3 + 12y^4$

4. $9a^4b - 18a^5b + 27a^6b$

5. $-3x^2 + 6x - 33$

6. $-12p^3q - 8p^2q^2 + 4pq^3$

7. $2w(x+3) - 5(x+3)$

8. $8y(a+b) + 9(a+b)$

Answers with explanations

1. $4(x) - 4(5) = 4(x-5)$

2. $3w(2w) + 3w(1) = 3w(2w+1)$

3. $3y^3(5) + 3y^3(4y) = 3y^3(5+4y)$

4. $9a^4b(1) - 9a^4b(2a) + 9a^4b(3a^2)$
   $= 9a^4b(1 - 2a + 3a^2)$

5. $-3(x^2) + (-3)(-2x) + (-3)(11)$
   $= -3(x^2 - 2x + 11)$

6. $-4pq(3p^2) + (-4pq)(2pq) + (-4pq)(-q^2)$
   $= -4pq(3p^2 + 2pq - q^2)$

7. $(x+3)(2w) - (x+3)(5)$
   $= (x+3)(2w-5)$

8. $(a+b)(8y) + (a+b)(9)$
   $= (a+b)(8y+9)$

Factoring By Grouping

This method of factoring is utilized when you have 4 terms in a polynomial like $3ax + 12a + 2bx + 8b$. The first step in any factoring problem is to see if there is a common factor throughout every term of the polynomial. After that has been explored, continue with the chosen method.

To factor by grouping, start by splitting the polynomial in half with the first two terms being grouped together $(3ax + 12a)$ and the second two terms being grouped together $(+ 2bx + 8b)$. *Note:* the sign in front of the 3rd term goes with the second grouping. Next, find the GCF of the first 2 terms and factor it out. For the example, $3a$ is the GCF. Factored, it leaves $3a(x + 4)$. Then find the GCF of the second two terms and factor it out. The GCF is $2b$. Factored, it is $2b(x + 4)$. Once you've completed this step, the remaining binomial of the two groups should be the same: $(x + 4)$. Factor this out as we did in the GCF section. $(x + 4)(3a + 2b)$.

8 examples for practice.

Factor by grouping.

1. $5x + 10y + ax + 2ay$
2. $ax + ay / - bx - by$
3. $tu - tv - 2u + 2v$
4. $6x^2 + 3x + 4x + 2$
5. $3j^2k + 15k + j^2 + 5$
6. $14w^6x^6 + 7w^6 - 2x^6 - 1$
7. $16w^4 - 40w^3 - 12w^2 + 30w$
8. $3ab^2 + 6b^2 - 12ab - 24b$

Answers with explanations:

1. Answer: $(x + 2y)(5 + a)$

   $5x + 10y \mid + ax + 2ay$
   $= 5(x + 2y) + a(x + 2y)$
   $= (x + 2y)(5 + a)$

2. Answer: $(x + y)(a - b)$

   $ax + ay \mid - bx - by$
   $= a(x + y) - b(x + y)$
   $= (x + y)(a - b)$

3. Answer: $(u - v)(t - 2)$

   $tu - tv \mid - 2u + 2v$
   $= t(u - v) - 2(u - v)$
   $= (u - v)(t - 2)$

4. Answer: $(2x + 1)(3x + 2)$

   $6x^2 + 3x \mid + 4x + 2$
   $= 3x(2x + 1) + 2(2x + 1)$
   $= (2x + 1)(3x + 2)$

5. Answer: $(j^2 + 5)(3k + 1)$

   $3j^2k + 15k \mid + j^2 + 5$
   $= 3k(j^2 + 5) + 1(j^2 + 5)$
   $= (j^2 + 5)(3k + 1)$

6. Answer: $(2x^6 + 1)(7w^6 - 1)$

   $14w^6x^6 + 7w^6 \mid -2x^6 - 1$
   $= 7w^6(2x^6 + 1) - 1(2x^6 + 1)$
   $= (2x^6 + 1)(7w^6 - 1)$

7. Answer: $2w(2w - 5)(4w^2 - 3)$

   $16w^4 - 40w^3 \mid -12w^2 + 30w$
   $= 2w(8w^3 - 20w^2 \mid -6w + 15)$
   $= 2w[4w^2(2w - 5) - 3(2w - 5)]$
   $= 2w(2w - 5)(4w^2 - 3)$

8. Answer: $3b(a + 2)(b - 4)$

   $3ab^2 + 6b^2 \mid -12ab - 24b$
   $= 3b(ab + 2b \mid -4a - 8)$
   $= 3b[b(a + 2) - 4(a + 2)]$
   $= 3b(a + 2)(b - 4)$

Factoring Trinomials In The Form of $x^2 + bx + c$

For these trinomials, notice that the coefficient of the first term is 1. That is crucial for utilizing this method.

The first step is to identify the factors of the c term. You need to find which factors of c have a sum equal to the b term. Here's an example: $x^2 + 8x + 15$. $b = 8$ and $c = 15$. The factors of 15 are 1 and 15, 3 and 5. The sums of these factors are 16 and 8, respectively. We need the factors of 15 whose sum is 8, the b term. This is 3 and 5. Therefore the factor of $x^2 + 8x + 15 = (x + 3)(x + 5)$.

Another example: $y^2 + 4y - 12$. b is 4 and c is -12. The factors are 1 and -12, -1 and 12, 2 and -6, -2 and 6, 3 and -4, and -3 and 4. The sums, respectively, are -11, 11, -4, 4, -1, and 1. For this problem, you need the factors of -12 whose sum is 4. So -2 and 6 work. The factored binomials are: $(y + 6)(y - 2)$.

8 problems for practice

Factor.

1. $x^2 + 4x - 45$

2. $w^2 - 15w + 50$

3. $p^2 - 8p - 48$

4. $w^2 - 5w - 6$

5. $10t^3 - 30t^2 - 40t$

6. $-2c^2 - 22cd - 60d^2$

7. $-x^2 + x + 12$

8. $-a^2 + 6a - 8$

Answers with explanations

1. $(x + 9)(x - 5)$

Factors	Sums
1 and -45	-44
-1 and 45	44
5 and -9	-4
-9 and 5	4

   *Because 4 is the sum you needed, the factors of 9 and -5 are the solution for this problem.

2. $(w - 10)(w - 5)$

Factors	Sums
1 and 50	51
-1 and -50	-51
2 and 25	27
-2 and -25	-27
5 and 10	15
-5 and -10	-15

   *You're looking for the sum of -15, which is the result of -5 and -10.

3. $(p - 12)(p + 4)$

Factors	Sums
1 and -48	-47
-1 and 48	47
2 and -24	-22
-2 and 24	22

3 and -16	-13
-3 and 16	13
4 and -12	-8
-4 and 12	8
6 and -8	-2
-6 and 8	2

*The sum -8 is what the problem requires. This is obtained by combining 4 and -12.

4. $(w + 1)(w - 6)$

Factors	Sums
1 and -6	-5
-1 and 6	5
2 and -3	-1
-2 and 3	1

*The sum of -5 results from 1 and -6.

5. $10t(t - 4)(t + 1)$

First you must factor out the GCF of $10t$, leaving $10t(t^2 - 3t - 4)$. The factors, 1 and -4, provide the sum of -3 for the middle term.

6. $-2(c + 5d)(c + 6d)$

First, factor out the GCF of -2, resulting in $-2(c^2 + 11cd + 30d^2)$. Then find the factors of $30d^2$ whose sum is 11. Those factors are $5d$ and $6d$. Don't forget to incorporate the $d's$ in the second terms of the binomials.

7. $-(x-4)(x+3)$

You must initially factor out the negative 1, leaving $-1(x^2 - x - 12)$. Factoring results in $-1(x-4)(x+3)$ based on the -4 and 3 being factors of -12 whose sum is -1, the b term.

8. $-(a-2)(a-4)$

First factor out -1. The result is $-(a^2 - 6a + 8)$. Using -2 and -4 as the factors whose sum is -6, you have $-(x-2)(x-4)$.

You should be feeling more confident about factoring at this point. Continue to practice. We have a few more types.

Factoring Trinomials of the form $Ax^2 + Bx + C$ by the AC-Method

In the previous section, all of the trinomials had a 1 as the leading coefficient. This method is used for trinomials that have an integer other than 1 as the leading coefficient. These polynomials are of the form of $Ax^2 + Bx + C$ where A is the coefficient of the x^2 term. B is the coefficient of the x term. C is the constant.

Steps:

1. Check for GCF of entire polynomial.
2. Multiply A and C.
3. List the factors of that product.
4. Determine which factors add up to B.
5. Rewrite the middle term using these factors.
6. Factor by grouping.

Let's use the example, $2x^2 + 9x + 4$, to show this method. $A = 2, B = 9, C = 4$. Multiplying $A \times C$: $2 \times 4 = 8$. The factors of 8 are 1 and 8, 2 and 4. The sums are 9 and 6, respectively. We want the term that equals B: 9. This would be the 1 and 8.

The rewrite of the middle term yields $2x^2 + 1x + 8x + 4$. Split the polynomial in preparation for factoring by grouping: $2x^2 + 1x | + 8x + 4$. The GCF of the first half is x. The GCF of the second half is 4. After factoring those GCF's out, you have $x(2x + 1) + 4(2x + 1)$. Factor out the common binomial, yielding $(2x + 1)(x + 4)$ which is the answer to the example.

8 practice problems

1. $2x^2 + 7x + 6$

2. $8x^2 - 2x - 3$

3. $10x^3 - 85x^2 + 105x$

4. $9y^3 - 30y^2 + 24y$

5. $-18x^2 + 21xy + 15y^2$

6. $-8x^2 - 8xy + 30y^2$

7. $2x^4 + 5x^2 + 2$

8. $2p^2 - 8p + 3$

Answers with explanations

1. Answer: $(2x + 3)(x + 2)$

$$A = 2, B = 7, C = 6$$

$A \times C = 2 \times 6 = 12$ Find the factors of 12 whose sum is $B(7)$

Factors of 12	Sums:
1 and 12	13
-1 and -12	-13
2 and 6	8
-2 and -6	-8
3 and 4	7
-3 and -4	-7

Once we identify the factors as 3 and 4, we rewrite the polynomial, changing the middle terms:

$$2x^2 + 3x + 4x + 6$$

Divide the polynomial in half and factor by grouping:
$$x(2x + 3) + 2(x + 3)$$

Factor out common term:
$$(2x + 3)(x + 2)$$

2. Answer: $(2x + 1)(4x - 3)$

$$A = 8, B = -2, C = -3$$

$A \times C = 8 \times (-3) = -24$ Find the factors of -16 whose sum is -2

Factors of -24	Sums:
1 and -24	-23
-1 and 24	23
2 and -12	-11
-2 and 12	11
3 and -8	-5
-3 and 8	5
4 and -6	-2
-4 and 6	2

Once we identify the factors as 4 and -6, we rewrite the polynomial, changing the middle terms:
$$8x^2 + 4x - 6x - 3$$

Divide the polynomial in half and factor by grouping:
$$4x(2x + 1) - 3(2x + 1)$$

Factor out common term:

$(2x + 1)(4x - 3)$

3. Answer: $5x(x - 7)(2x - 3)$

There is a GCF that you must factor out first: 5x

$5x(2x^2 - 17x + 21)$
Now $A = 2, B = -17, and\ C = 21$

$A \times C = 42$

Factors	Sums
1 and 42	43
-1 and -42	-43
2 and 21	23
-2 and -21	-23
3 and 14	17
-3 and -14	-17
6 and 7	13
-6 and -7	-13

The -3 and -14 are the factors you need. Now rewrite with the new middle terms.

$5x(2x^2 - 3x - 14x + 21)$

Factor by grouping.

$5x[x(2x - 3) - 7(2x - 3)]$

4. Answer: $3y(3y-4)(y-2)$

GCF: $3y$

$$3y(3y^2 - 10y + 8)$$

$$A = 3, B = -10, C = 8$$

$$A \times C = 24$$

Factors	Sums
1 and 24	25
-1 and -24	-25
2 and 12	14
-2 and -12	-14
3 and 8	11
-3 and -8	-11
4 and 6	10
-4 and -6	-10

Rewrite:

$$3y(3y^2 - 6y - 4y + 8)$$

Factor by grouping.

$$3y[3y(y-2) - 4(y-2)]$$

5. Answer: $-3(2x + y)(3x - 5y)$

Factor out GCF: $-3(6x^2 - 7xy - 5y^2)$

$$A \times C = 6 \times (-5) = -30$$

24

Factors	Sums
1 and -30	-29
-1 and 30	29
2 and -15	-13
-2 and 15	13
3 and -10	-7
-3 and 10	7
5 and -6	-1
-5 and 6	1

3 and -10 are the factors you need for the sum of -7.

Rewrite: $-3(6x^2 + 3xy - 10xy - 5y^2)$

Factor by grouping:

$-3[3x(2x + y) - 5y(2x + y)]$

6. Answer: $-2(2x - 3y)(2x + 5y)$

Factor out GCF:

$-2(4x^2 + 4xy - 15y^2)$

$A \times C = 4 \times 15 = 60$

Factors	Sums
1 and -60	-59
-1 and 60	59
2 and -30	-28
-2 and 30	28
3 and -20	-17
-3 and 20	17
4 and -15	-11

-4 and 15	11
5 and -12	-7
-5 and 12	7
6 and -10	-4
-6 and 10	4

Rewrite:

$-2(4x^2 - 6xy + 10xy - 15y^2)$

Factor by grouping:

$-2[2x(2x - 3y) + 5y(2x - 3y)]$

7. Answer: $(2x^2 + 1)(x^2 + 2)$

Same concept using higher ordered variables.

$A \times C = 4$

Factors	Sums
1 and 4	5
-1 and -4	-5
2 and 2	4
-2 and -2	-4

Using 1 and 4, rewrite:

$2x^4 + x^2 + 4x^2 + 2$

Factor by grouping:

$x^2(2x^2 + 1) + 2(x^2 + 1)$

8. Prime.

$A \times C = 2 \times 3 = 6$

Factors	Sums
1 and 6	7
-1 and -6	-7
2 and 3	5
-2 and -3	-5

**As you can see none of the factors of 6 have the sum of -8. This means that this polynomial is PRIME and cannot be factored.

Factoring Using the Trial-and-Error Method

This method is another way of factoring trinomials that are of the form $Ax^2 + Bx + C$. As always, first look for a common factor for the entire trinomial and factor that out. Then, list the pairs of factors of the A term and the pairs of factors of the C term. Your objective is to combine these factors to find the B term as a sum. This is done by inputting the pairs of factors of the A term in the front blanks (?) of the binomials and inputting the pairs of factors of the C term in the back blanks (%) of the binomials. (?x + %)(?x + %). Then multiply the outer 2 terms and the inner 2 terms. Add them together to find the combination that equals the B term.

8 practice problems

1. $10x^2 - 9x - 1$

2. $3b^2 + 8b + 4$

3. $8y^2 + 13y - 6$

4. $40x^3 - 104x^2 + 10x$

5. $-45x^2 - 3xy + 18y^2$

6. $6w^2 - 25w + 4$

7. $-4x^2 + 26xy - 40y^2$

8. $3x^4 + 8x^2 + 5$

Answers with explanations

1. Answer: $(10x + 1)(x - 1)$
The factors of 10 are 1 and 10, 2 and 5. The factors of -1 are 1 and -1.
Now you have to combine them in various combinations to see which has the sum of $-9x$.
Try: 1 and 10 as the front factors and 1 and -1 as the back factors. $(x + 1)(10x - 1)$

Outer $x \times (-1) = -1x$
Inner $1 \times 10x = 10x$

Adding the products: $-1x + 10x = 9x$

We are looking for $-9x$. It looks like we have the correct factors, but the opposite sign. Let's switch the signs in the binomials. The result:
$(x - 1)(10x + 1)$

2. Answer: $(3b + 2)(b + 2)$
The factors of 3 are 1 and 3. The factors of 4 are 1 and 4, 2 and 2.
Inserting the front factors:
$(b + \_)(3b + \_)$
Inserting the back factors:
$(b + 1)(3b + 4)$
Outer product: $4 \times b = 4b$;
Inner product: $1 \times 3b = 3b$
Sum of the products: $4b + 3b = 7b$
You are looking for $8b$ so the factors need to be changed. Exchange the 1 and 4.

Try again. $(b + 4)(3b + 1)$
Outer products: $1b$
Inner products: $4 \times 3b = 12b$
Sum of the products: $1b + 12b = 13b$
This is still not what we need. So let's use 2 and 2.
$(b + 2)(3b + 2)$
Outer products: $2 \times b = 2b$
Inner products: $2 \times 3b = 6b$
Sum of products: $2b + 6b = 8b$
This is the correct result.

3. Answer: $(y + 2)(8y - 3)$
Factors of 8: 1 and 8, 2 and 4
Factors of -6: 1 and -6, -1 and 6, 2 and -3, -2 and 3
You are looking for the combination of factors whose outer product and inner product sum is $13y$.
Start with 1 and 8 as the front products and 1 and -6 as the back products.

$(y + 1)(8y - 6)$

The sum of outer and inner products is $-6y + 8y = 2y$, which is too small

Changing the back factors around: $(y - 6)(8y + 1)$, the outer and inner product sum is $y - 48y = -47y$ which is too large, number-wise.

Change the rear factors to 2 and -3: $(y + 2)(8y - 3)$
The sum of inner and outer products =
$-3y + 16y = 13y$. Bingo!
The answer is: $(y + 2)(8y - 3)$

4. Answer: $2x(2x - 5)(10x - 1)$
Remember, you always have to look for the GCF first. $2x$ is the GCF of this trinomial: $2x(20x^2 - 52x + 5)$

Now, factor the remaining trinomial.
The factors of 20: 1 and 20, 2 and 10, 4 and 5
The factors of 5: 1 and 5.
Because you are looking for a negative middle number, -52, you have to also consider the negatives of all of the above listed factors. So the factors of 20 also include: -1 and -20, -2 and -10, -4 and -5. Along with the factors of 5 being -1 and -5.

You can use the negative rear factors and try 1 and 20 as the front factors:

$(x-1)(20x-5)$

Outer and inner products sum: $-5x - 20x = -25x$

Using 2 and 10 as front factors: $(2x-1)(10x-5)$

Outer and inner products sum: $-10x - 10x = -20x$

Because this isn't correct, keep trying by changing the placement of the rear factors: $(2x-5)(10x-1)$

Outer and inner products sum: $-2x - 50x = -52x$

Final answer: $2x(2x-5)(10x-1)$

5. Answer: $-3(3x+2y)(5x-3y)$

GCF is -3. Factored out, it leaves $-3(15x^2 + xy - 6y^2)$

Factors of $15x^2$: $1x$ and $15x$, $3x$ and $5x$

Factors of $-6y^2$:

$-1y$ and $6y$, $1y$ and $-6y$, $-2y$ and $3y$, $2y$ and $-3y$

We need factors that lead to the inner sum of 1 so we can pick ones that will yield inner and outer products that are closer together.

Try: 3 and 5 as front factors and -2 and 3 as back factors to see what results.

$(3x-2y)(5x+3y)$

Outer and inner product sums: $9xy - 10xy = -xy$

We have the correct amount, but the incorrect sign, so let's change the signs inside the binomials.

$(3x+2y)(5x-3y)$

The final answer is

$-3(3x+2y)(5x-3y)$

6. Answer: $(6w - 1)(w - 4)$

The factors of 6: 1 and 6, 2 and 3
The factors of 4: 1 and 4, -1 and -4, 2 and 2, -2 and -2
Try: $(6w - 1)(w - 4)$
Outer product: $-24w$
Inner product: $-w$
Sum of products: $-25w$ *This is the one.

7. Answer: $-2(2x - 5)(x - 4y)$

First factor out the GCF: $-2(2x^2 - 13xy + 20y^2)$
Factors of 2: 1 and 2
Factors of 20: 1 and 20, 2 and 10, 4 and 5, as well as their negatives
Try $(2x - 2y)(x - 10y)$
Outer product: $-20xy$
Inner product: $-2xy$
Sum of products: $-22xy$ *too large
Try $(2x - 4y)(x - 5y)$
Outer product: $-10xy$
Inner product: $-4xy$
Sum of products: $-14xy$ *almost there
Try switching the back terms. $(2x - 5y)(x - 4y)$
Outer product: $-8xy$
Inner product: $-5xy$
Sum of products: $-13xy$ *YES!

8. Answer: $(3x^2 + 5)(x^2 + 1)$

This one is simple because 3 and 5 only have one set of factors. So you just have to figure out which place to put each factor in.

$(3x^2 + 1)(x^2 + 5)$ yields $16x^2$ as the sum of products. Since you're looking for $8x^2$, you have to exchange the 1 and 5 to $(3x^2 + 5)(x^2 + 1)$.

Difference Of Squares and Perfect Square Trinomials

The difference of squares is a special polynomial. If you start with $(x + y)(x - y)$ and multiply it out, the result is $x^2 - y^2$. The difference of squares is factoring the $x^2 - y^2$ for the result of $(x + y)(x - y)$.

To factor, these polynomials, first look for the GCF. Remember that's always your first step. After factoring out the GCF, if there is one, determine the square root of both terms. The factor is written in this form: The quantity of the square root of the 1st term plus the square root of the 2nd term multiplied by the difference of the square root of the 1st term and the square root of the 2nd term. $(?x + \%y)(?x - \%y)$. Example: $9x^2 - 25y^2$. The square root of the first term is $3x$. The square root of the second term is $5y$. The factored form of $9x^2 - 25y^2 = (3x + 5y)(3x - 5y)$.

This is called the difference of squares because the sign between the two binomials is subtraction. This is not the same as the sum of squares, such as $x^2 + y^2$. The sum of squares is PRIME and cannot be factored.

8 practice problems

1. $y^2 - 25$

2. $a^2 - 64$

3. $49s^2 - 4t^2$

4. $16q^2 - 81w^2$

5. $18w^2z - 2z$

6. $w^4 - 16$

7. $y^3 - 5y^2 - 4y + 20$

8. $p^2 + 9$

Answers with explanations

1. $(y + 5)(y - 5)$
Square root of y^2: y
Square root of 25: 5

2. $(a + 8)(a - 8)$
Square root of a^2: a
Square root of 64: 8

3. $(7s + 2t)(7s - 2t)$
Square root of $49s^2$: $7s$
Square root of $4t^2$: $2t$

4. $(4q + 9w)(4q - 9w)$
Square root of $16q^2$: $4q$

Square root of $81w^2$: $9w$

5. $2z(3w + 1)(3w - 1)$
Factor out the GCF: $2z$
   $2z(9w^2 - 1)$
Square root of $9w^2$: $3w$
Square root of 1: 1

6. $(w^2 + 9)(w + 3)(w - 3)$
Square root of w^4: w^2
Square root of 81: 9
Answer: $(w^2 + 9)(w^2 - 9)$
*Because $w^2 - 9$ is another difference of squares, it must be factored again.
Square root of w^2: w
Square root of 9: 3
Factored: $(w + 3)(w - 3)$

7. $(y - 5)(y + 2)(y - 2)$
Factor by grouping: $y^2(y - 5) - 4(y - 5)$
   $(y - 5)(y^2 - 4)$

$(y^2 - 4)$ is the difference of squares, also. Factored, you get $(y + 2)(y - 2)$

8. $p^2 + 9$ is PRIME and cannot be factored.

Perfect Square Trinomials

Perfect square trinomials are the result of squaring binomials.

$(a + b)^2 = (a + b)(a + b) = a^2 + 2ab + b^2$

$(a - b)^2 = (a - b)(a - b) = a^2 - 2ab + b^2$

For example: $(3x + 5)^2 = (3x)^2 + 2(3x)(5) + (5)^2$

$= 9x^2 + 30x + 25$

To see if the polynomial is a perfect square trinomial, follow these 2 steps:

Step 1: Determine if the first and third terms are both perfect squares and have positive coefficients. If so, they are a and b, respectively.

Step 2: If step 1 is true, determine if the middle term is two times the product of a
and b to equal $2ab$ or $-2ab$.

8 Practice Problems

1. $x^2 + 6x + 9$

2. $y^2 - 10y + 25$

3. $x^2 + 14x + 49$

4. $x^2 - 6x + 9$

5. $25y^2 - 20y + 4$

6. $81w^2 + 72w + 16$

7. $18c^3 - 48c^2d + 32cd^2$

8. $5w^2 + 50w + 45$

Answers with explanations

1. Answer: $(x + 3)^2$
Square root of x^2? x
Square root of 9? 3
Middle term $2ab$ or $-2ab$? $2(3)(x) = 6x$
Both criteria are met, therefore this is a perfect square trinomial.

2. Answer: $(y - 5)^2$
Square root of y^2? y
Square root of 25? 5
Middle term $2ab$ or $-2ab$? $-2(5)(y) = -10y$

3. Answer: $(x + 7)^2$
Square root of x^2? x
Square root of 49? 7
Middle term $2ab$ or $-2ab$? $2(7)(x) = 14x$

4. Answer: $(x - 3)^2$
Square root of x^2? x
Square root of 9? 3
Middle term $2ab$ or $-2ab$? $-2(3)x = -6x$

5. Answer: $(5y - 2)^2$
Square root of $25y^2$? $5y$
Square root of 4? 2
Middle term $2ab$ or $-2ab$? $-2(2)(5y) = -20y$

6. Answer: $(9w - 4)^2$
Square root of $81w^2$? $9w$
Square root of 16? 4

Middle term $2ab$ or $-2ab$? $2(4)(9w) = 72w$

7. Answer: $2c(3c - 4d)^2$
Factor our GCF: $2c(9c^2 - 24cd + 16d^2)$
Square root of $9c^2$? $3c$
Square root of $16d^2$? $4d$
Middle term $2ab$ or $-2ab$? $-2(3c)(4d) = -24cd$

8. Answer: $5(w + 9)(w + 1)$
Factor our GCF: $5(w^2 + 10w + 9)$
Square root of w^2? w
Square root of 9? 3
Middle term $2ab$ or $-2ab$? $2(3)(w) = 6w$ No! The middle term of the original problem is $10w$. This problem is not a perfect square trinomial, but can be factored by a different method.

Sum and Difference of Cubes

Another set of special and more advanced cases for factoring are the sum or difference of cubes. These are generated from expanding

$(a + b)(a^2 - ab + b^2)$ and

$(a - b)(a^2 + ab + b^2)$.

The steps in determining if you're dealing with the difference or sum of cubes, follow these 2 steps.

Step 1: Does the first term have a cube root?
Step 2: Does the second term have a cube root?

If the answer is yes to both questions, then you can utilize the formulas (listed below) to factor.

The formulas are:

$$a^3 + b^3 = (a + b)(a^2 - ab + b^2)$$

$$a^3 - b^3 = (a - b)(a^2 + ab + b^2)$$

These formulas will be derived in a higher math course, but for now the best thing to do is memorize them.

As an example, when factoring $x^3 - 8$, identify the cube root of x^3 as x. This is your a term. Then identify the cube root of 8 as 2. That is the b term.

Then fill in the terms of the formula:

$$(x - 2)(x^2 + 2x + 2^2)$$

$$= (x - 2)(x^2 + 2x + 4)$$

8 Practice Problems

1. $x^3 + y^3$

2. $y^3 + 8$

3. $27y^3 - 1$

4. $x^3 y^6 - 64$

5. $8x^3 - 27y^3$

6. $64u^3 + 125v^3$

7. $2y^4 - 16yz^3$

8. $27x^3 + 343y^3$

Answers with explanations

1. Cube root of x^3: x
   Cube root of y^3: y
   Sum of cubes: $(x + y)(x^2 - xy + y^2)$

2. Cube root of y^3: y
   Cube root of 8: 2
   Sum of cubes: $(y + 2)(y^2 - 2y + 4)$

3. Cube root of $27y^3$: $3y$
   Cube root of 1: 1
   Difference of cubes: $(3y - 1)(9y^2 + 3y + 1)$

4. Cube root of x^3y^6: xy^2
   Cube root of 64: 4
   Difference of cubes: $(xy^2 - 4)(x^2y^4 + 4xy^2 + 16)$

5. Cube root of $8x^3$: $2x$
   Cube root of $27y^3$: $3y$
   Difference of cubes: $(2x - 3y)(4x^2 + 6xy + 9y^2)$

6. Cube root of $64u^3$: $4u$
   Cube root of $125v^3$: $5v$
   Sum of cubes: $(4u + 5v)(16u^2 - 20uv + 25v^2)$

7. Factor out the GCF: $2y(y^3 - 8z^3)$
   Cube root of y^3: y

Cube root of $8z^3$: $2z$
Difference of cubes: $2y(y - 2z)(y^2 + 2yz + 4z^2)$

8. Cube root of $27x^3$: $3x$
Cube root of $343y^3$: $7y$
Sum of cubes: $(3x + 7y)(9x^2 - 21xy + 49y^2)$

In Closing

This book has been a basic, but comprehensive, look at all types of factoring. Did you get everything you needed in 8 minutes? I hope so. If it took you a little longer to fully grasp everything, that's alright, too. If you've followed the steps and practiced the problems provided, you should be able to factor any type of problem with ease and agility.

Other Titles by Arlissa Pinkelton:

Learn About Exponents in 8 Minutes

Learn All You Need To Know About Real Numbers in 8 Minutes

Learn Radicals in 8 Minutes

Learn Rational Expressions in 8 Minutes

Learn To Add, Subtract, Multiply and Divide Polynomials in 8 Minutes

Solving & Graphing Inequalities in 8 Minutes

Solving & Graphing Linear Equations in 8 Minutes

www.ingramcontent.com/pod-product-compliance
Lightning Source LLC
Chambersburg PA
CBHW070719180526
45167CB00004B/1534